A Maker's Guide to SCREWS

Written by JOHN WOOD

Illustrated by AMY LI

LERNER PUBLICATIONS ◆ MINNEAPOLIS

First American edition published in 2021 by Lerner Publishing Group, Inc.

Editor: Madeline Tyler
Design: Drue Rintoul

Lerner Publications Company
An imprint of Lerner Publishing Group, Inc.
241 First Avenue North
Minneapolis, MN 55401 USA

For reading levels and more information, look up this title at www.lernerbooks.com.

Main body text set in VAG Rounded Std. Typeface provided by Adobe Systems.

Library of Congress Cataloging-in-Publication Data

Names: Wood, John, 1990- author. | Li, Amy, illustrator.
Title: A maker's guide to screws / written by John Wood ; illustrated by Amy Li.
Other titles: Screws
Description: First American edition. | Minneapolis : Lerner Publications, 2021. | Series: A maker's guide to simple machines | Includes index. | Audience: Ages 6–10 | Audience: Grades 2–3 | Summary: "From lightbulbs and bottle caps to water taps and peanut butter lids, screws are everywhere! Learn about screws and make your own spinning top with the help of an alien engineer"— Provided by publisher.
Identifiers: LCCN 2020054269 (print) | LCCN 2020054270 (ebook) | ISBN 9781728416441 (lib. bdg.) | ISBN 9781728438337 (pbk.) | ISBN 9781728418896 (eb pdf)
Subjects: LCSH: Screws—Juvenile literature.
Classification: LCC TJ1338 .W66 2021 (print) | LCC TJ1338 (ebook) | DDC 621.8/82—dc23

LC record available at https://lccn.loc.gov/2020054269
LC ebook record available at https://lccn.loc.gov/2020054270

Manufactured in the United States of America
1-48891-49205-12/17/2020

Words that look like <u>this</u> can be found in the glossary on page 24.

BE A MAKER

Look at this huge mess. This is Maker. He does this all the time. Say sorry, Maker.

Believe it or not, Maker is very smart when it comes to machines.

A machine is an object that makes a job easier to do. Maker wants to teach you about one of the simplest types of machine: a screw.

SCREWS

Screws are long and thin with a ridge running around the outside. They are usually used to hold two things together. Although they look a lot like each other, screws are different from nails.

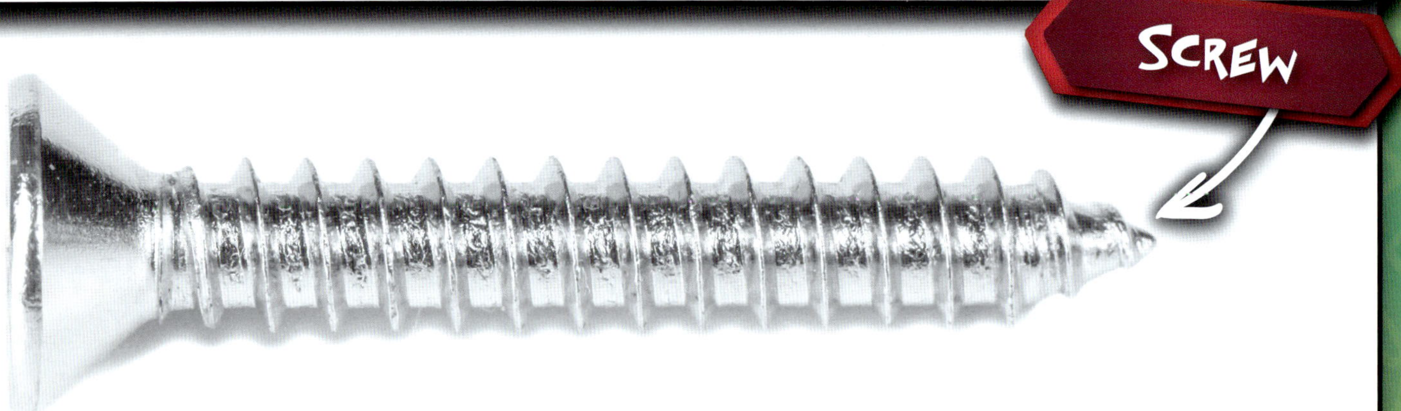

Nails do not have ridges, and they are hit into an object with a hammer. Screws need a twisting motion to force them into an object. Wrenches, drills, or screwdrivers are used for screws.

There are lots of different screws. Here are some screws that we see every day.

The caps of some pens are screwed on and off.

The tuning pegs of some musical instruments are screws.

Some light bulbs have a screw at the bottom.

Plastic bottle caps are a type of screw.

Faucets often have a screw that controls the flow of water.

There are four main parts of a screw: the head, the tip, the shaft, and the ridges that run around the outside. The ridges are called threads.

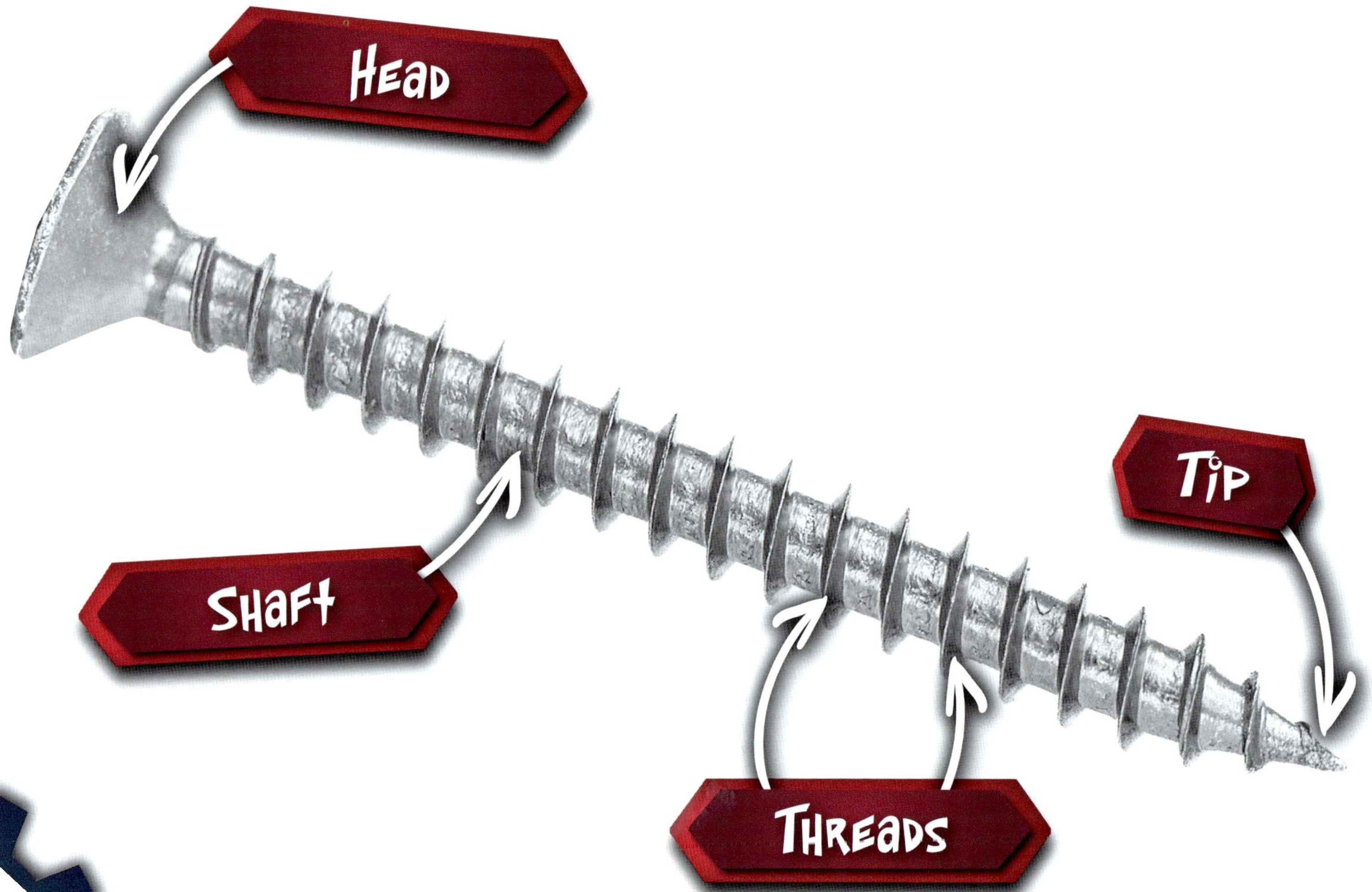

The tip is pointy and sharp. This helps the screw to go into another object. The head is wide and flat. This is the part that a screwdriver or drill fits into.

Screws like these are usually made out of metal.

There are lots of different types of screws. Maker wants to show you a few with different heads. These screws need a wrench or a screwdriver with the right end.

These are common shapes for screw heads.

A bolt is like a thicker screw, and it comes with a nut. The nut is a piece of metal that screws onto the tip of the bolt after it has been driven through something.

The nut makes the bolt more secure.

A screw turns a twisting force into a downward force. As the screw turns, it sinks into a material such as wood or plastic.

The threads help to keep the screw in place. The threads cause extra friction so the screw won't fall out. This makes it good at holding things together.

Some screws can be huge. These bolts are holding a pipe together.

The less space there is between the threads, the less force is needed to screw it. However, it will take more turns. The more space there is, the more force is needed.

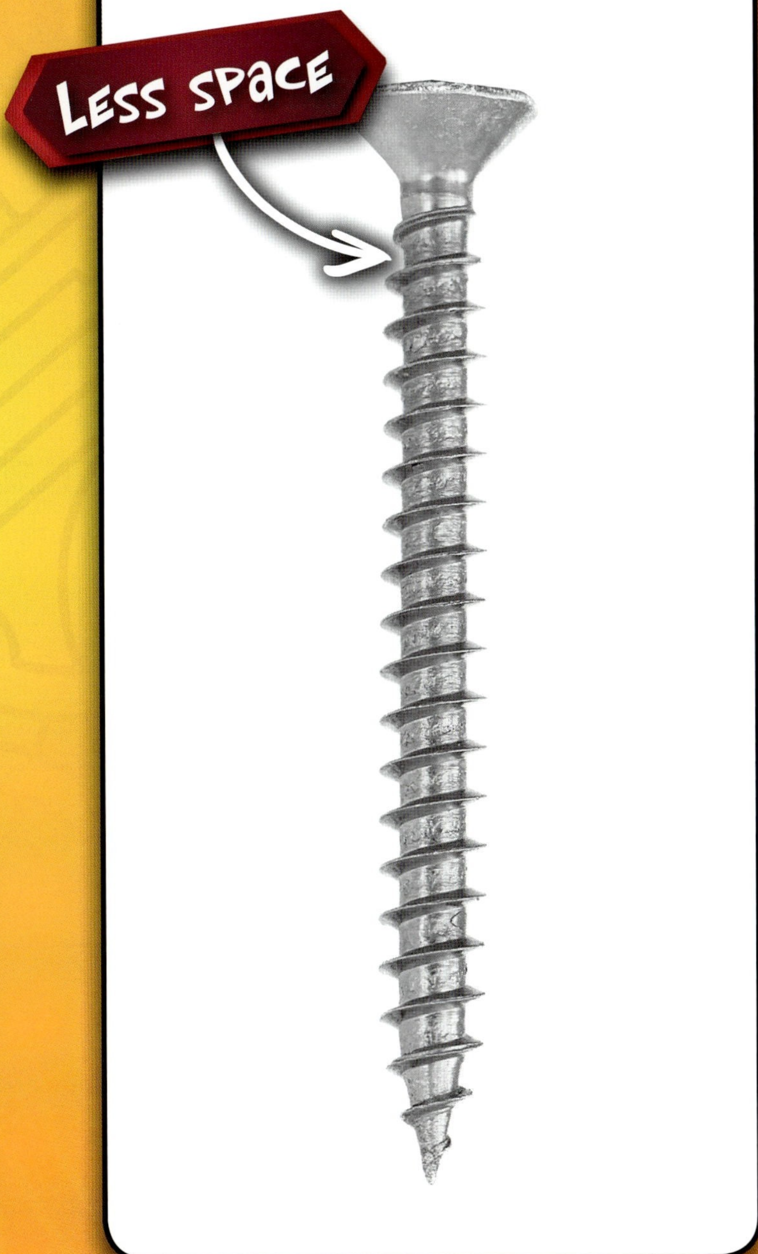

BUILD A SPINNING TOP

It is time to build! We will be using a screw to make a spinning top. Get an adult to help you with the drilling in step 1.

The toy will also use a magnet.

- A long, thin screw
- 2 nuts that fit the screw
- 2 light wooden disks
- A small magnet that will fit on the head of the screw

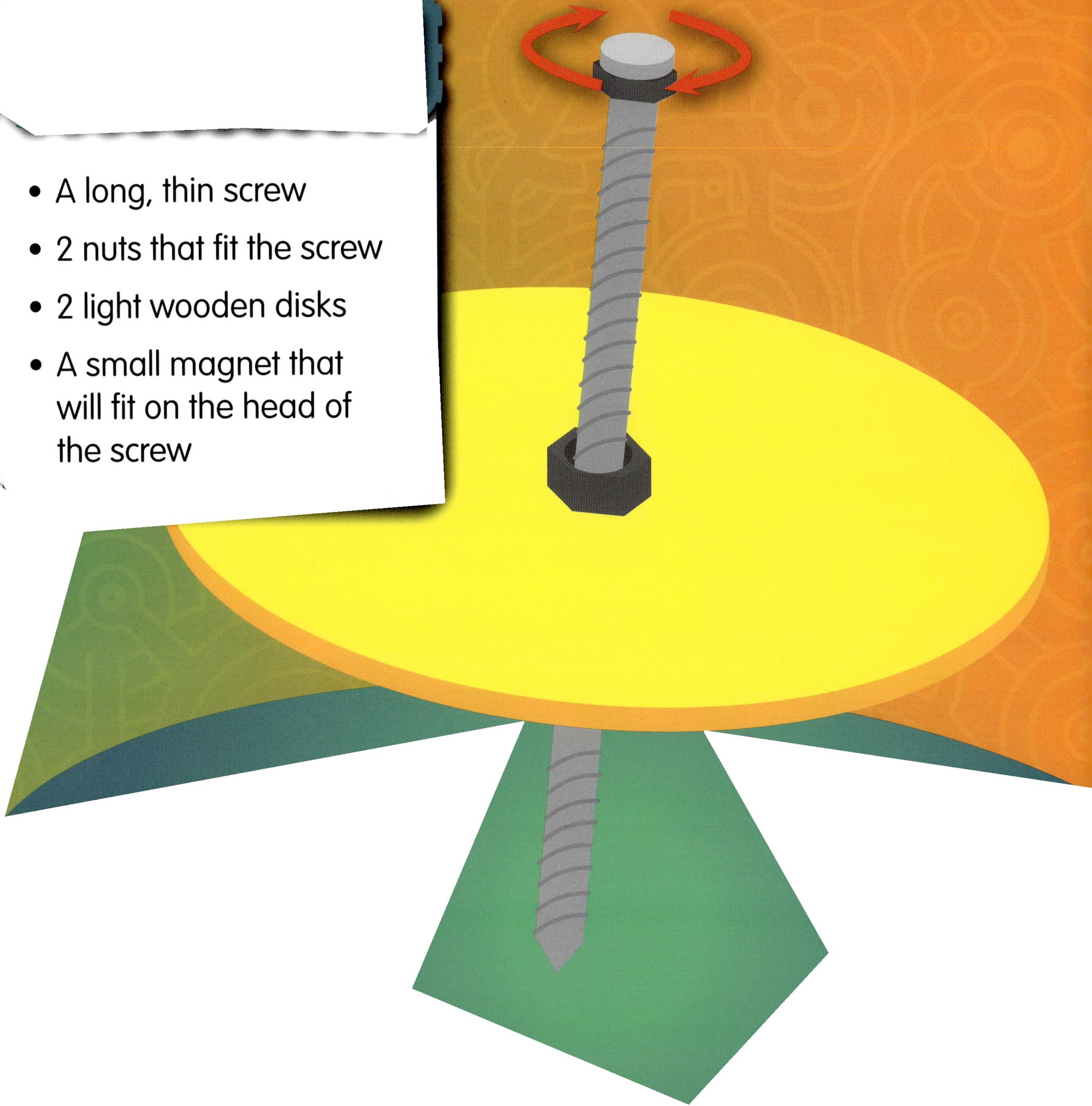

STEP 1

Cut or drill a hole in the middle of both wooden disks.

STEP 2

Put a nut on the screw and screw it about halfway up the shaft.

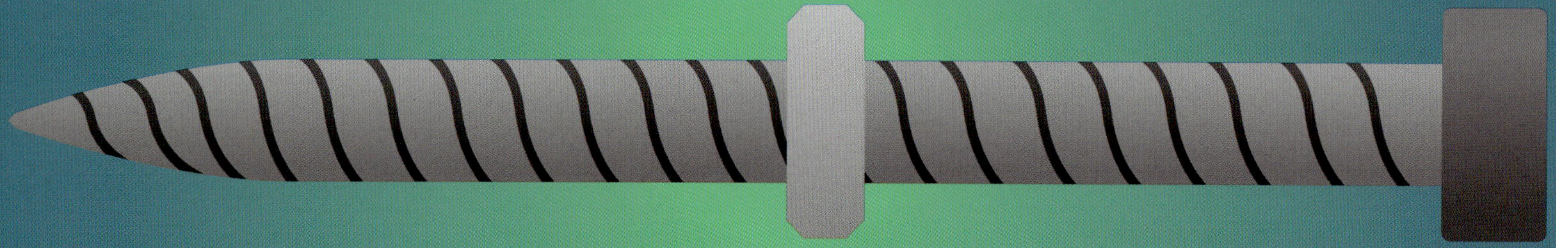

STEP 3

Now stick the screw through the middle of both wooden disks.

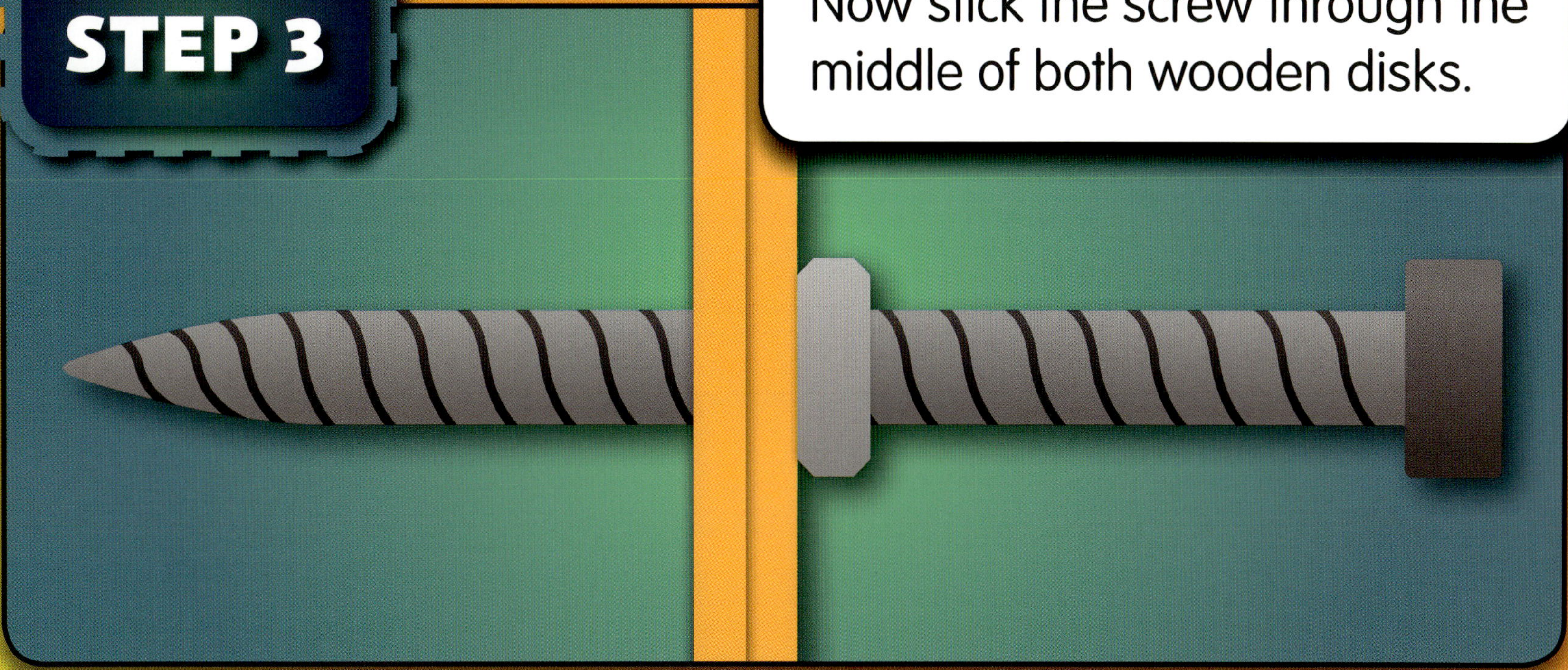

STEP 4

Put the other nut on the screw and screw it up so it touches the wooden disks.

STEP 5

Make sure the nuts are tight and the wooden disks are firmly in place.

STEP 6

Stick the magnet to the head of the screw.

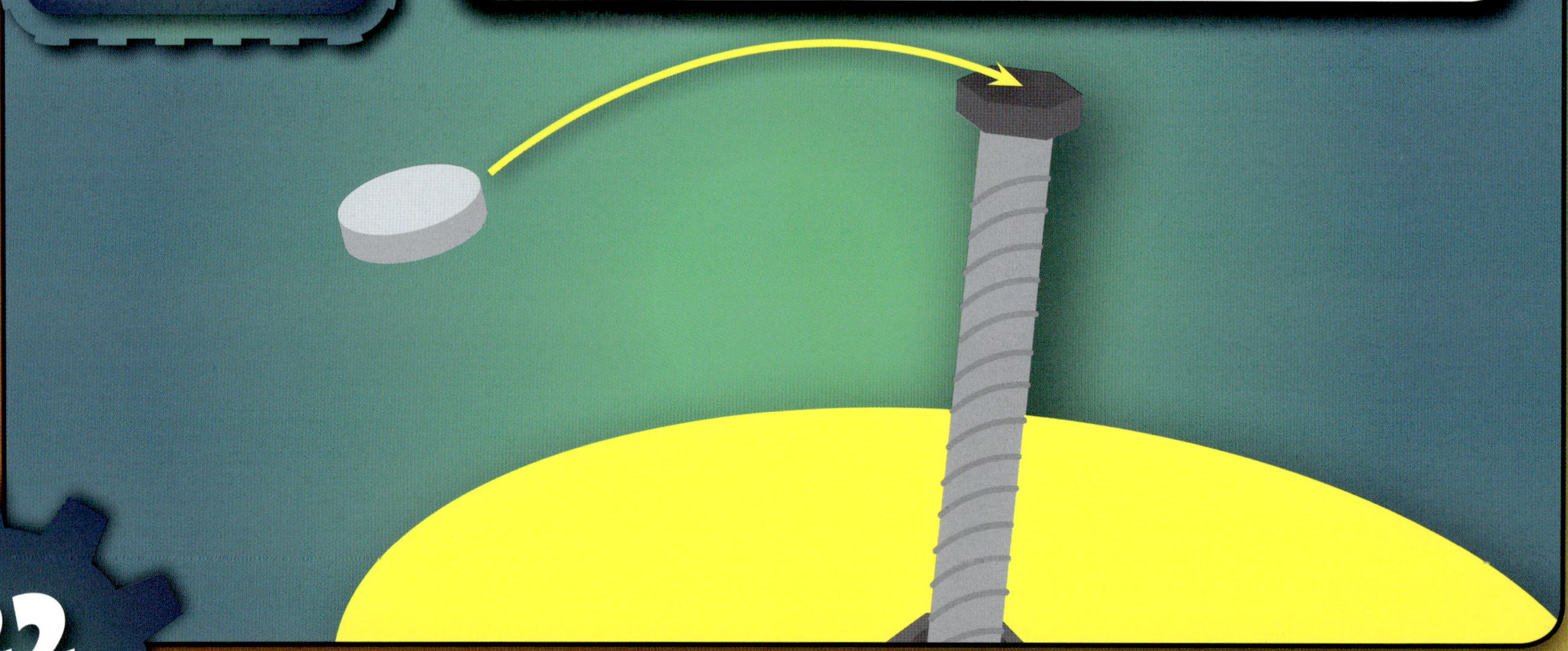

STEP 7

Hold the toy by the head and spin it!

STEP 8

You could bend some thin metal wire into different shapes and spin the toy over it. See what happens!

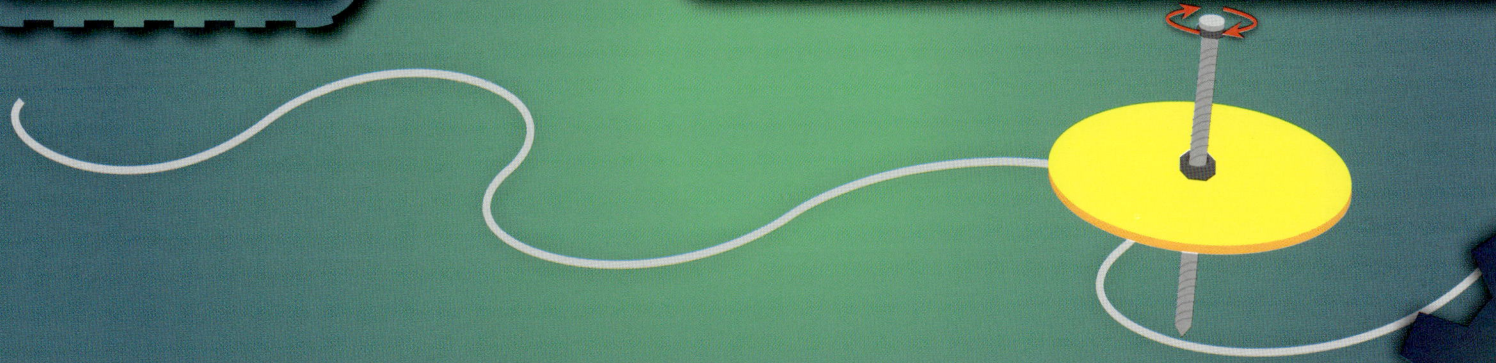

force	a push or pull on an object
friction	a force that slows things down and creates heat when two things rub together
material	something from which objects are made
motion	movement
ridge	a raised edge
secure	safe and strong

INDEX